AF228145

# INVERTEBRATE GROUPS

BY CAROL HAND

Cover Image: Coral colonies can be found in reefs around the Andaman Islands, which are located east of India and southwest of Myanmar.

Core Library

An Imprint of Abdo Publishing
abdobooks.com

Cover Photo: Shutterstock Images
Interior Photos: Kate Besler/Shutterstock Images, 4–5; Leo Kohout/Shutterstock Images, 8, 43; Shutterstock Images, 10, 20–21, 25, 33, 36, 45; Ihor Hvozdetskyi/Shutterstock Images, 12–13; Westend61/Getty Images, 15; Red Line Editorial, 16; Pavel Krasensky/Shutterstock Images, 23; Lea McQuillan/500px/Getty Images, 28–29; Craig Holdcroft/Shutterstock Images, 31; Gerald Corsi/iStockphoto, 34–35; Anthony Pierce/Alamy, 39; Anton Petrus/Moment/Getty Images, 40

Editor: Laura Stickney
Series Designer: Ryan Gale

**Library of Congress Control Number: 2024948991**

**Publisher's Cataloging-in-Publication Data**

Names: Hand, Carol, author.
Title: Invertebrate groups / by Carol Hand
Description: Minneapolis, Minnesota: Abdo Publishing, 2026 | Series: Strength in numbers: animal groups | Includes online resources and index.
Identifiers: ISBN 9781098297275 (lib. bdg.) | ISBN 9798384919797 (ebook)
Subjects: LCSH: Invertebrates--Juvenile literature. | Invertebrate communities--Juvenile literature. | Animal colonies--Juvenile literature. | Invertebrates--Behavior--Juvenile literature.
Classification: DDC 592--dc23

# CONTENTS

# MIGRATING MONARCHS

t's late spring in the eastern United States. The air is warm, and flowers are blooming. An adult monarch butterfly has just emerged from her chrysalis. She flies around her garden home feeding on flower nectar. She mates with a male monarch. Then the butterfly finds a milkweed plant. She lays each egg on its own milkweed leaf and secures it with a drop of glue that she makes inside her body. During the next two to five weeks, the butterfly lays 300 to 500 eggs.

Monarch butterflies have bright-orange wings with black lines and white spots. A group of monarchs is sometimes called a flutter, a kaleidoscope, or a swarm.

The butterfly's eggs hatch into caterpillars, which feed on the milkweed. It is the only food they can eat. For two weeks, the caterpillars eat, grow, and molt. Then they each form a chrysalis in which they undergo metamorphosis. During this process, the caterpillars develop into adults. Later, each one emerges from its chrysalis as a butterfly.

During the summer, several generations of monarchs undergo this life cycle. Adults live for only two to six weeks. In late summer or early fall, the days get shorter and cooler. The monarchs' behavior changes.

## THREATS TO MONARCHS

Eastern monarch butterfly populations fell by 59 percent during the 2023–2024 season. Herbicides are weed-killing chemicals. In the United States, the use of herbicides destroys the milkweed plants on which monarch caterpillars feed. In Mexico, agriculture and tourism threaten monarchs' forest habitats. Climate change disrupts monarch migration too. It changes weather patterns in overwintering areas and in summer breeding grounds.

The last summer generation of monarchs can live up to nine months. This generation does not reproduce. Instead, these butterflies gather in groups. Together, the butterflies migrate to central Mexico, where it is warmer. They spend the winter there. Some butterflies migrate as far as 3,000 miles (4,830 km). Monarchs survive this long journey by flying in air currents and thermals.



Thermals are rising currents of warm air. Doing this allows the butterflies to glide quickly through the air without flapping their wings as much. This way, they use less energy.

In Mexico, millions of clustered monarchs cover the trunks and branches of trees. Many monarch overwintering sites are part of the Monarch Butterfly Biosphere Reserve.

The monarchs arrive in Mexico in late October or early November. They cluster together on the branches of oyamel fir trees. Clustering helps the butterflies stay warm. They are mostly inactive in the winter but sometimes fly on warm days. During this time, the butterflies need water but not food. They remain in Mexico until March, when they migrate north again.

# WHAT ARE INVERTEBRATES?

Monarchs are one kind of invertebrate that sometimes forms large groups. Invertebrates are animals that do not have backbones. Many animals, such as dogs, elephants, and birds, are vertebrates. They have backbones. But about 90 to 97 percent of all animals on Earth are invertebrates.

A large portion of invertebrate species are insects. Some invertebrates, such as insects, spiders, and worms, are largely land-based. Other invertebrate species live underwater. Ocean invertebrates include crustaceans, such as lobsters, crabs, and krill. Corals, mollusks, oysters, and clams are other types of ocean invertebrates.

# INVERTEBRATE SOCIETIES

Some invertebrates live in large, organized groups called societies. Societies help species survive. All individuals in the society work together. In many invertebrate societies, there is a division of labor.

This means each member does its own job. Together,
they help keep the group healthy and safe.

Invertebrates may work together to build structures
such as termite mounds. These animals often work
together to find food too. In groups, they can gather
more food, search larger areas, and share meals. Living
in societies also helps animals escape predators and
protect their young. This makes it easier for animals to
find mates, reproduce, and care for their babies.

Most land-based invertebrate societies are
formed by insects. These include ants, bees, termites,
and wasps. Insect societies may contain thousands

or even millions of members. Each society includes individuals with different body types. These individuals do specific jobs in the group.

Some ocean-based invertebrates are sessile, or unmoving. These individuals live attached to the ocean floor and to each other. They form groups called colonies. Examples include corals and sponges. Other ocean invertebrate societies are free-living. Many are plankton. These organisms float with ocean currents and tides. Krill are one plankton group. They gather in huge swarms, providing a major food source for fish, whales, and penguins. Forming groups helps many different invertebrates survive and thrive in the wild.

## EXPLORE ONLINE

Chapter One talks about invertebrates. The article at the website below goes into more depth on this topic. Does the article answer any of the questions you had about invertebrates?

### INVERTEBRATE

abdocorelibrary.com/invertebrate-groups

# HONEYBEES

Honeybees are important insects. They are pollinators. They carry pollen from plant to plant, which fertilizes the plants. This allows the plants to reproduce by creating seeds or fruit. Today, honeybees pollinate about 80 percent of the world's flowering plants.

Honeybees are also social insects. They live and work together in colonies. They have several features that make them social. First, al honeybees work together to care for their

Honeybees have many small hairs on their bodies. Pollen sticks to these hairs and then falls off as bees visit different flowers and plants.

colony's offspring. Second, generations in the colony overlap. Offspring assist their parents. This means the colony continues when one generation changes to the next.

## SOCIAL STRUCTURES

Honeybee colonies include three castes of bees. These are drones, workers, and queens.

Drones are males. They mate with queens from other colonies. By doing so, they pass the colony's genes to the next generation.

Worker bees are the smallest honeybees. They are females that do not reproduce. Instead, they carry out

many important jobs in the colony. They collect pollen and nectar to make honey. They care for eggs and young. They maintain and defend the hive too.

A colony's queen bee produces all the offspring. A queen can lay up to 2,000 eggs per day, which is more than her body weight. She can lay eggs almost constantly. Most honeybee colonies include one queen, as many as 60,000 worker bees, and a few hundred drones. Together, these bees build and live among wax combs in the hive. Combs provide storage space for honey and pollen. Bees can also make nests for their eggs and young in the combs.

A honeybee colony's population declines in the fall and winter. The queen lays many eggs during this time.

# HONEYBEE AGE VS. TASK

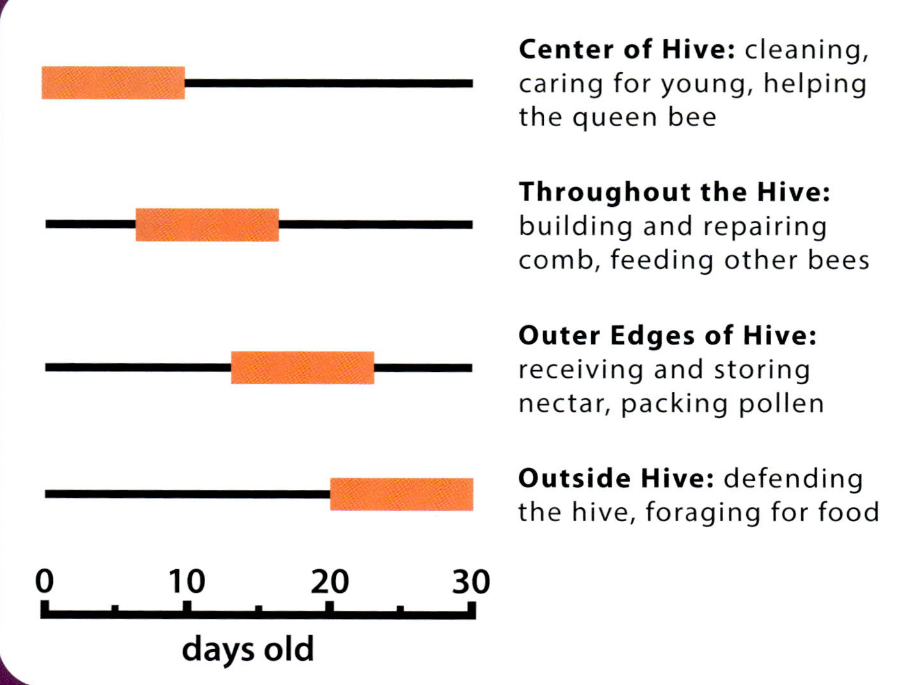



Worker bees take on new jobs as they grow older. This chart shows worker bees' locations and tasks within the hive at different ages after hatching. What do you notice about the different tasks in the chart? Why do you think a worker bee's job changes as it gets older?

The amount of pollen stored during the summer determines how many young bees will be raised.

In spring and summer, honeybees emerge from the hive to collect pollen and nectar. The colony expands. The queen lays eggs that will hatch into drones.

When a colony becomes too large for its hive, the bees swarm. Swarming is the process by which a group of bees, including the queen, gather to leave a colony and form a new one. Before leaving the hive, worker bees build queen cells. The queen lays eggs in these cells that will become new queen bees. Then the queen and about half of the worker bees leave the nest. This usually happens in spring. Most of the bees rest on a nearby tree for 24 to 36 hours while worker bees locate a new nest site. Then the swarm moves to the new site and builds another hive. This provides more space for honeybees. A new queen bee takes over the old colony.

## SUPERORGANISMS

A honeybee colony is considered a superorganism. This means the colony's members act as one large organism. Within the colony, each bee has its own job. Individual bees act like cells in an animal do. They carry out specialized tasks based on their role to keep the colony alive.

Honeybees work together to perform three major jobs. First, they reproduce the colony by swarming. Second, the bees keep the hive's temperature at around 93 degrees Fahrenheit (34°C). If the hive gets too hot, bees cool it by fanning water droplets with their wings. When the hive is too cold, bees warm it by vibrating their wings to produce heat. Third, bees work together to move air inside the hive. The bees' nesting cavities have little air flow. Bees gather at the hive entrance and fan air in and out.

## WASP COLONIES

Some types of wasps, such as yellowjackets and paper wasps, are social insects that live in colonies. Like honeybee colonies, wasp colonies are made up of mostly workers. These wasps gather food, care for the nest, and raise young. The colony also includes a queen wasp, who lays all the eggs. Like honeybee hives, wasp nests have combs. But some wasp nests look grayish and papery. Other times, wasps build their nests in the ground.

# STRAIGHT TO THE
# SOURCE

Locusts are insects related to grasshoppers. Like honeybees, they form swarms at specific times in their lives. A *National Geographic* article discusses locust swarms:

> *Locusts . . . form enormous swarms that spread across regions, devouring crops and leaving serious agricultural damage in their wake. . . . During dry spells, solitary locusts are forced together in the patchy areas of land with remaining vegetation. This sudden crowding . . . makes locusts more sociable and promotes rapid movements and more varied appetite. When rains return— producing moist soil and abundant green plants . . . locusts begin to produce rapidly and become even more crowded together. In these circumstances, they shift completely from their solitary lifestyle to a group lifestyle.*

> Source: "Locusts." *National Geographic*, n.d., nationalgeographic.com. Accessed 30 Oct. 2024.

## WHAT'S THE BIG IDEA?

Take a close look at this passage. What is the main connection being made between locust swarms and group behavior? How does swarming together help locusts survive?

# ANTS

There are more than 15,000 known species of ants on Earth. Experts believe there are likely many more. Ants are known for working together in large groups. These invertebrates also have a strong, organized social structure.

Ants live together in colonies of varying sizes. Some, such as those of clonal raider ants, number only 16 individuals. But army ant colonies in South American rainforests can have between 150,000 and 2 million individuals.

Worker ants take on different jobs as they grow older. Older workers often support their colonies by foraging for food.

# ARMY ANTS

Army ants live in Central America, South America, and Africa. They form swarms of up to 200,000 ants. They feed on spiders and crickets. Groups of army ants form protective nests called bivouacs using their own bodies. The ants use their limbs to hold onto each other, forming a shelter. They can quickly move these nests to other locations. More than 500 species follow army ant swarms to feed on their leftover prey. These include birds, lizards, and butterflies.

All ant colonies have a division of labor. Every ant has a job to do. Jobs might include caring for the nest or searching for food. Each ant doesn't know what other ants are doing. It doesn't know what the colony needs to survive. But when each ant does its job, the entire colony functions as an organized unit. In large colonies, the division of labor is more complex. There are more castes and jobs. The colony's large size protects the ants. By working together, the ants can quickly complete tasks in order to survive. They can also swarm together to defend their colony from predators.

## ANT LIFE CYCLES

Like honeybee colonies, most ant colonies have three castes. One caste is a winged queen. Another is made up of a few male drones. The third caste includes female worker ants that do not reproduce. These ants do jobs that help the colony survive, such as foraging for food.

As an ant colony grows, the queen lays some eggs that will become new queens. Shortly after a new queen hatches, she grows wings, leaves the nest, and mates with a drone from another colony. The queen stores sperm from this mating. After mating, the queen digs a nest and forms a new ant colony. She uses the stored sperm to fertilize eggs throughout her life. These fertilized eggs become worker ants and queen ants. The queen stays in the nest until she dies. During this time, she does nothing but lay eggs.

Ants begin as eggs. They develop into larvae and then pupae. Pupae hatch into adults. Unfertilized eggs hatch into drone ants.



## WORKING TOGETHER

Ants can accomplish big tasks by working

Leaf-cutter ants work as a team to harvest leaves. Some ants cut up leaves, while others help carry the leaf pieces back to their nest.

together as a group. Leaf-cutter ants in South American rainforests forage for leaves. When one leaf-cutter ant finds a good area of leaves, she creates a trail of chemical scents called pheromones. These act as a signal to other ants, guiding them to the spot. The ants work together to cut up leaves into pieces that are small enough to carry. Then they carry the leaf pieces to their nests. There, the ants maintain fungus gardens.

They grow this fungi as a food source. The leaves provide food for the fungi, helping it grow.

Living in groups also helps some ants catch prey. Some species, such as acacia ants and trap-jaw ants, are predators. They live on plants and eat prey that lands on them. One type of Amazonian tree-dwelling ant builds traps on plants. The ants use materials such as fungi to create a platform with holes in it. Then they hide inside the trap. When an insect lands on the trap, many ants ambush and grab the insect. By using these traps, the ants can capture prey much larger than themselves.

Other ants are raiders. They gather to attack nearby ant colonies and eat their larvae. Sometimes, the raiders capture and raise the other colony's larvae. When the larvae grow into adults, the raiders enslave them and make them workers. Other ants, such as the bullet ant, produce toxins. The ant can deliver a painful sting. This helps it defend itself and kill prey.

# STRAIGHT TO THE SOURCE

Deborah M. Gordon is a professor at Stanford University who studies ant societies. She wrote about how ants in a colony switch tasks:

> *We know now that ants do not perform as [specialized] factory workers. Instead ants switch tasks. An ant's role changes as it grows older and as changing conditions shift the colony's needs. . . . Yet in an ant colony, no one is in charge or tells another what to do. So what determines which ant does which task, and when ants switch roles? . . . An ant's social role is a response to interactions with other ants. In brief encounters, ants use their antennae to smell one another, or to detect a chemical that another ant has recently deposited. . . . These simple interactions between ants allow colonies to adjust the numbers performing each task and to respond to the changing world.*

Source: Deborah M. Gordon. "The Queen Does Not Rule." *Aeon*, 19 Dec. 2016, aeon.co. Accessed 30 Oct. 2024.

## CONSIDER YOUR AUDIENCE

Adapt this passage for a different audience, such as your principal or friends. Write a blog post conveying this same information for the new audience. How does your post differ from the original text and why?

# CORALS

Corals are underwater invertebrates. They are sessile organisms. Each coral is made up of hundreds of thousands of tiny organisms called polyps. Coral polyps have three cell layers. The inner layer lines the polyp's gut. The gut opening, or mouth, is surrounded by tentacles. These take in plankton for food, remove debris, and protect the polyp. A polyp's tentacles have stinging cells, which shoot toxic barbs.

The Great Barrier Reef, located off the coast of Australia, is home to about 600 different coral species. This includes hard and soft corals of all shapes, sizes, and colors.

# TYPES OF CORAL REEFS

The three major types of coral reefs are fringing reefs, barrier reefs, and atolls. Fringing reefs are the most common. They form close to the shore along edges of continents. They are separated from the shore by shallow lagoons. Barrier reefs have wider, deeper lagoons that separate them from the shore. Some corals in barrier reefs stick out of the water, forming a barrier between the lagoon and the open ocean. Atolls surround a lagoon where a coral island has sunk. The top of the island is usually the top of an underwater volcano. Atolls are ring shaped.

The polyp uses these to protect itself and capture prey.

Corals that build coral reefs are called stony corals. A coral reef is a colony of corals. They occur in shallow waters on both sides of the equator. Reefs cover less than one percent of Earth's surface, but are home to about one-fourth of all marine species.

## CORALS AND ALGAE

Coral polyps work with zooxanthellae, a

type of green algae. Algae are plantlike organisms that mostly grow in water. Polyps give the algae a place to live. The algae make food using photosynthesis. They give up to 90 percent of their food to the polyps. This cooperation is known as symbiosis. It helps both species survive and thrive.

However, by 2050, 90 percent of coral reefs could die because of climate change. Climate change causes ocean temperatures to rise. This causes coral bleaching. Corals' zooxanthellae produce toxins at high temperatures. When this happens, corals expel the zooxanthellae. They turn white and lose their main food source. Many corals die because of coral bleaching.

# CORAL COLONIES

Corals form when polyps release eggs and sperm into the water. The fertilized eggs form floating larvae. When a polyp matures, it attaches to a rock on the ocean floor. It begins to bud. It produces clones, or exact copies of itself. Each tiny polyp produces calcium carbonate, or limestone. Limestone from the polyps forms a skeleton. This joins the polyps together. Soon, the limestone forms a reef that weighs several tons. The limestone is secreted at the base of the polyps, and coral tissue forms a skin around the reef's top.

In coral colonies, there is no division of labor. Instead, all polyps function in

# PARTS OF A CORAL

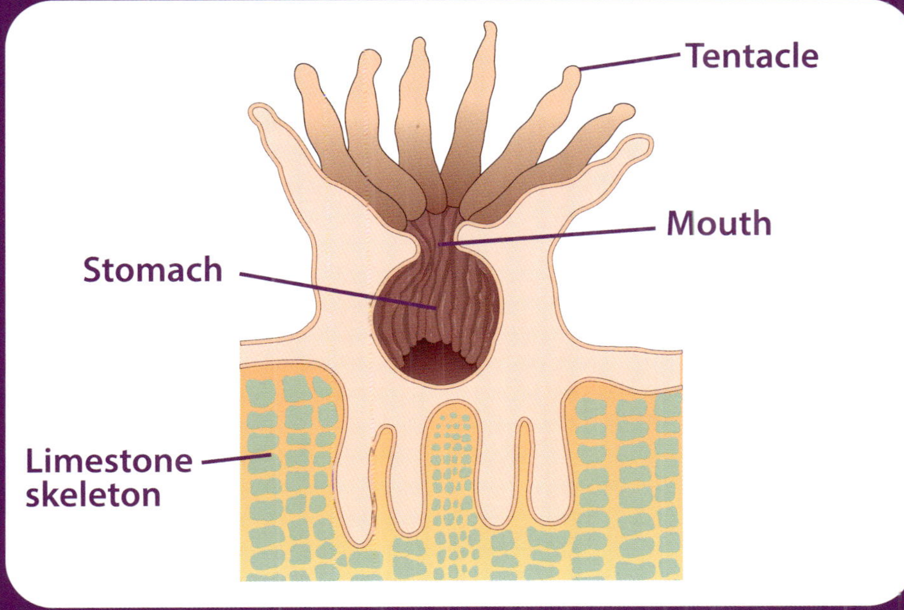



Coral polyps produce limestone, which forms a skeleton that holds the polyps together. This diagram shows a living polyp on top and the limestone skeleton beneath. What do you notice about the different parts of the coral?

the same way. Over time, corals connect with others nearby. Each coral depends on the others to survive. By forming colonies, they can spread over a larger area and reproduce. Polyps can also share nutrients from food with each other through their connective tissue. Corals in colonies work closely together. They form complex ecosystems that are like vast underwater cities.

# KRILL

**K**rill are a type of plankton. Plankton are any organisms that do not usually swim but are carried along on ocean waves and tides. Ocean plankton are so tiny that many of them can't be seen. But these small creatures are essential to life on Earth.

Phytoplankton, or plantlike plankton, form the base of the ocean's food webs. They live and photosynthesize in sunny regions of oceans. This is usually in the

Antarctic krill are one of five krill species that live in the ocean near Antarctica. They have large black eyes and pinkish, translucent bodies.

top 330 feet (100 m) of the water. Zooplankton, or animal-like plankton, live among phytoplankton. Some zooplankton spend their entire lives floating in oceans. Corals are considered plankton only when in the larval stage. But then they sink, attach to the ocean floor, and become sessile. Once this happens, they are no longer plankton.

More than 30,000 species of zooplankton live in swarms throughout the oceans. They feed on phytoplankton, other zooplankton, and organic waste. They are also important prey for many animals.

Krill are one of the largest zooplankton. About 85 species of zooplankton are krill. These tiny, shrimplike crustaceans average 2.5 inches (6.4 cm) in length, about the size of a human's pinky finger. Krill have hard outer skeletons and many legs.

Unlike most plankton, krill are excellent swimmers. They make daily vertical migrations in huge swarms, swimming up toward the ocean's surface and then back down. During the day, they swim hundreds of meters deep to avoid predators. At night, the krill return to the water's surface to feed.

## LIGHT SHRIMP

Antarctic krill are sometimes called light shrimp. This is because of bioluminescent, or light-producing, organs on their undersides. These organs are known as photophores. To predators looking up, the soft glow of the photophores makes the krill blend in with the sunlight shining down on the water's surface. Light production may also help krill communicate and maintain group structure. Deeper in the ocean, where there is no light, it may also help with survival, communication, and predator–prey interactions.

Krill swim 2 feet (0.6 m) per second. Swimming in a group helps the krill cover long distances. As they swim, the krill avoid predators by swimming backward and rapidly flipping their back ends. This helps them swim quickly.

## KRILL SWARMS

Krill maintain huge swarms through their own individual behavior and through social interactions. Their behavior controls the swarm's arrangement. It also controls the swarm's direction and how much it moves.

Humpback whales use a method called lunge feeding to catch krill. They push themselves upward through a krill swarm with open mouths. Some krill get trapped in the whales' baleen.

Individual krill can use currents created by other krill in the swarm to save energy as they swim. This helps the group swim faster and farther than a single krill can.

Krill swarms can be miles wide and miles deep. They are sometimes large enough to be seen from space. These massive swarms are a major food source for humpback whales, gray whales, and blue whales.

Many seabirds, including puffins, catch and eat krill.

The whales use their mouths to take in massive amounts of water containing krill. Then they filter out the krill using their baleen plates. These hairlike bristles hang from a whale's upper mouth. Seals, squid, penguins, and many birds and fish also eat krill. However, swimming in a large group can help protect krill from these predators. It makes it harder for predators to single out and catch any one krill.

Some krill species, such as Antarctic krill, occur in larger groupings called patches, ranging from very small swarms to gigantic super-swarms. These super-swarms may be several miles long. Being part of a super-swarm can make it easier and safer for krill to find food. It can also help krill find mates and communicate with others in the swarm.

Krill patches vary in size, sex, and maturity. A patch of krill may also contain many swarms that touch each other but do not merge. By grouping together in this way, more krill can survive. This group behavior is just one of the ways that invertebrates such as krill thrive.

## FURTHER EVIDENCE

Chapter Five discusses the importance of krill swarms in the ocean. What is the main point of this chapter? What key evidence supports this point? Read the article at the website below. Does the information on the website support the main point of the chapter? Does it present new evidence?

## THRILL TO THE KRILL

abdocorelibrary.com/invertebrate-groups

# FAST FACTS

- Many invertebrates, or animals without backbones, form groups to survive.

- Monarch butterflies gather in large groups to migrate. They travel south to overwinter in Mexico, where they cluster together on trees to stay warm. They migrate north to reproduce in the spring and summer.

- Honeybees and ants form colonies that have a division of labor. All individuals cooperate to care for young, collect and store food, and keep the colony alive.

- Most honeybee and ant colonies include three castes. These are female workers, male drones, and a queen. Workers do not reproduce. They perform specific jobs within the colony.

- Honeybees form swarms. The queen and half of the worker bees leave a colony to form a new colony.

- Coral reefs consist of many individual coral polyps held together by a limestone exoskeleton built by the polyps themselves. There is no division of labor among corals.

- Krill are shrimplike ocean plankton that form huge swarms. These swarms swim down during the day to avoid predators. At night, they swim up to the water's surface to feed.

- Krill swarms can be miles wide and miles deep. The swarms are a major food source for whales, seals, squid, and penguins.

### Tell the Tale

Chapter One of this book discusses monarch butterfly migration. Imagine you are watching monarchs migrate and arrive at an overwintering site in Mexico. Write 200 words about what you see. How do the butterflies work together?

### Take a Stand

Many invertebrate species that form groups are threatened by factors such as habitat loss and climate change. Some people are working to protect these species. Do you think people should do more to help threatened invertebrates? Why or why not?

### Why Do I Care?

You may not see many honeybee colonies where you live. But that doesn't mean you can't think about why honeybee colonies are important. How do honeybees work together? How do they affect the ecosystem? How might honeybees affect the food you eat and the plants around you?

## You Are There

This book discusses coral reefs. Imagine you are snorkeling in a coral reef. Write a letter home telling your friends about what you see there. What do the corals in the reef look like? What kinds of plants and animals do you see? Be sure to add plenty of detail to your notes.

# GLOSSARY

**caste**
a group of insects that look the same and carry out specific tasks

**chrysalis**
the protective covering in which a caterpillar develops into an adult butterfly

**climate change**
a process that is causing global temperatures to rise

**fertilize**
to join male and female cells to produce young

**gene**
a unit of information in DNA that controls traits passed down from parents to offspring

**molt**
to shed and replace an outer covering, such as skin or an exoskeleton

**organism**
a living plant or animal

**pesticide**
a chemical used to kill pests such as insects or weeds

**photosynthesis**
the process by which organisms such as plants turn sunlight, water, and carbon dioxide into food

**pollen**
a powder produced by plants that spreads to and fertilizes other plants

# ONLINE RESOURCES

To learn more about invertebrate groups, visit our free resource websites below.

Visit **abdocorelibrary.com** or scan this QR code for free Common Core resources for teachers and students, including vetted activities, multimedia, and booklinks, for deeper subject comprehension.

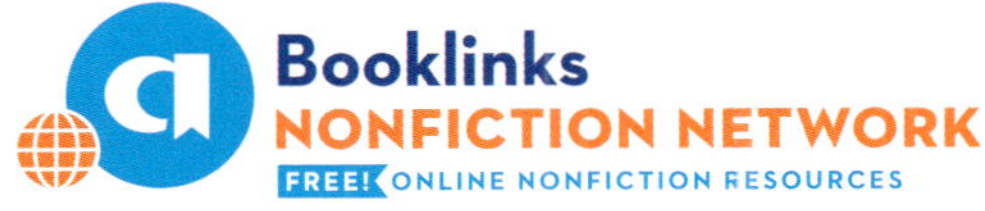

Visit **abdobooklinks.com** or scan this QR code for free additional online weblinks for further learning. These links are routinely monitored and updated to provide the most current information available.

# LEARN MORE

Lane, Laura. *Butterflies*. Abdo, 2023.

Mills, Andrea. *Oceans*. DK, 2020.

Mound, Laurence. *Insect*. DK, 2023.

# INDEX



## About the Author

Carol Hand has a PhD in zoology with a specialization in marine ecology and a special interest in environmental and climate science. She has taught college classes, written for standardized testing companies, and developed multimedia science curricula. She has also written more than 60 educational books for young people. More recently, she has written three books in a young adult science fiction series.